COMMON FISH FARMING CALCULATIONS

- A LAYMAN'S GUIDE

Koushik Roy

ICAR-JRF, SRF, NET and ARS (Mains) Qualified
M.F.Sc. Aquaculture (First class 1st)
B.Sc. Industrial Fish and Fisheries (First class 1st)

Dr. M.S. Chari

Professor, Dept. of Fisheries, College of Agriculture, I.G.K.V. Raipur, Chhattisgarh

Dr. S.R. Gaur

Professor and Head, Dept. of Fisheries, College of Agriculture, I.G.K.V. Raipur, Chhattisgarh

ISBN: 1511653019
ISBN-13: 978-1511653015

First Edition.
CreateSpace Publications, USA (An Amazon Associate)

DEDICATION

Dedicated to My Parents Mr. Ashok Kumar Roy and Mrs. Provati Roy; My
Sister Miss Saswati Roy for being my home and happiness.
Also to My professors Dr. Satyam Kumar Kundu, Dr. S.R. Gaur and Dr.
M.S. Chari for making me who I am today. I Salute You All.

CONTENTS

ACKNOWLEDGMENTS

I pay my heartfelt gratitude to all my juniors and acquainted fish farmers who urged on releasing this book as soon as possible to serve their purpose at various levels.

1 DETERMINING POND AREAS

Very few ponds are perfectly shaped squares, rectangles, triangles or circles. Common area formulas may need a little correction.

1. For Square/ Rectangular Shaped Ponds

If a pond is a perfect square or rectangle, the following formula applies:

A = l x w

Where A = Area (in square meter)

l = length in meter

w = width in meter

Example #1.1. Measurements show a regular rectangular pond to have dimensions of 120 m length and 80 m breadth. What is the surface area?

120 m

80 m

A = l x w

A = 120 x 80

A = 9600square meter (m²).

Convert it into hectares/acre by the following formula:

Area (in hectares) = Area (in square meter) /10000

= 9600/ 10000 = 0.96 ha (hectares).

OR, Area (in acres) = Area (in square meter) /4048

= 9600/ 4048 = 2.37 acres.

*NOTE: 1 hectare (ha) land is equal to 10000 m² i.e. - a pond which is 100 m in length and 100 m in breadth. 1 ha area is equivalent to 2.47 acres. 1 acre land is equal to 4048 m².

Example #1.2. An irregular rectangular pond has the following dimensions. What is the area?

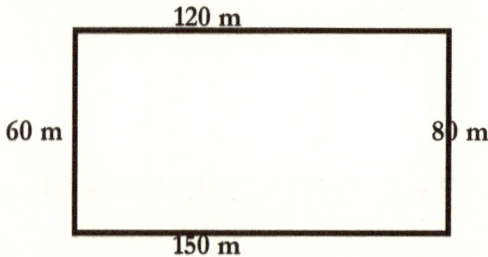

Area = [(l₁ + l₂)/ 2] x [(w₁ + w₂)/ 2]

= [(120+150)/ 2] x [(60 + 80)/ 2]

= (270/ 2) x (140/ 2)

= 135 x 70

= 9450 square meter (m²)

Convert it into hectares/acre by the following formula:

Area (in hectares) = Area (in square meter) / 10000

= 9450/ 10000 = 0.945 ha (hectares).

OR, Area (in acres) = Area (in square meter) / 4048

= 9450/ 4048 = 2.33 acres.

*NOTE: 1 hectare (ha) land is equal to 10000 m² i.e. - a pond which is 100 m in length and 100 m in breadth. 1 ha area is equivalent to 2.47 acres. 1 acre land is equal to 4048 m².

2. For Square/ Rectangular Shaped Ponds

Calculating pond areas for triangular-shaped ponds is easy. Just use the following formula:

Area = (a x b)/ 2

Where, a = short side (in meter)

b = long side (in meter)

The above formula used for calculating area works well for any type of triangular pond.

Example #1.3. Calculate the area of a triangular pond with the dimension given below.

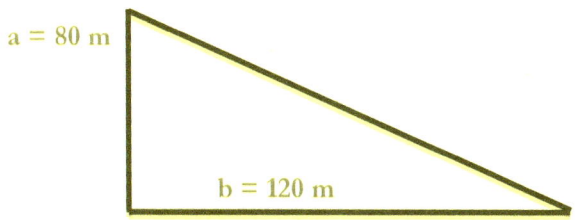

Area = (a x b)/ 2
= (80 x 120)/ 2
= 9600/ 2
= 4800 square meter (m²)

Convert it into hectares/acre by the following formula:

Area (in hectares) = Area (in square meter) / 10000
= 4800/ 10000 = 0.48 ha (hectares).

OR, Area (in acres) = Area (in square meter) / 4048
= 4800/ 4048 = 1.19 acres.

⁺NOTE: 1 hectare (ha) land is equal to 10000 m² i.e. - a pond which is 100 m in length and 100 m in breadth. 1 ha area is equivalent to 2.47 acres. 1 acre land is equal to 4048 m².

3. For Trapezoidal Shaped Ponds

To calculate areas for trapezoid-shaped ponds having four sides and a nearly or exactly 90 degree angle, use the following formula:

Area = [(a + b) x h]/ 2

Where,

Example #1.4. Calculate the area of a trapezoid shaped pond having the dimensions given in the figure below.

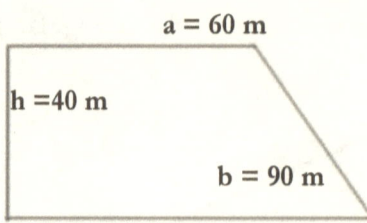

a = 60 m

h = 40 m

b = 90 m

Area = [(a + b) x h]/ 2
= [(60 + 90) x 40]/ 2
= [150 x 40]/ 2
= 6000/ 2
= 3000 square meter (m²)
<u>Convert it into hectares/acre by the following formula:</u>
Area (in hectares) = Area (in square meter) / 10000
= 3000/ 10000 = 0.3 ha (hectares).
OR, Area (in acres) = Area (in square meter) / 4048
= 3000/ 4048 = 0.74 acres.
*NOTE: 1 hectare (ha) land is equal to 10000 m² i.e. - a pond which is 100 m in length and 100 m in breadth. 1 ha area is equivalent to 2.47 acres. 1 acre land is equal to 4048 m².

4.   <u>For Circular Ponds</u>
If a pond is circular or near circular, the following formula applies:
Area = 3.14 x (d/ 2)²
Where, d = Diameter of the pond (in meters)

Example #1.5. Calculate the area of a circular shaped pond having the diameter of 80 meter.

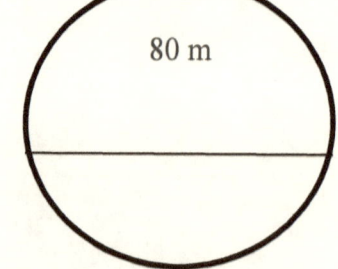

80 m

Area = 3.14 x (d/ 2)²
= 3.14 x (80/ 2)²
= 3.14 x 40²
= 3.14 x 1600
= 5024 square meter (m²)
<u>Convert it into hectares/acre by the following formula:</u>

Area (in hectares) = Area (in square meter) / 10000
= 5024/ 10000 = 0.5 ha (hectares).
OR, Area (in acres) = Area (in square meter) / 4048
= 5024/ 4048 = 1.24 acres.
*NOTE: 1 hectare (ha) land is equal to 10000 m² i.e. - a pond which is 100 m in length and 100 m in breadth. 1 ha area is equivalent to 2.47 acres. 1 acre land is equal to 4048 m².

5. For Irregularly Shaped Ponds

For calculating areas for irregularly-shaped or mixed- shaped ponds, divide the pond into regular shaped sections and calculate those areas. Then add the areas of each section to determine the total area for the pond.

2 DETERMINING AVERAGE DEPTH OF PONDS

Calculating the volume of ponds requires an accurate estimate of the pond's average depth. Two people are needed to calculate the depth of a pond – one to measure depth and one to record the measurements.

For ponds of 5 acres or less, a minimum of ten measurements is needed. For ponds over 5 acres, a minimum of 20 measurements must be taken. Take the measurements along an S-shaped figure across the pond. Refer to the figure below:

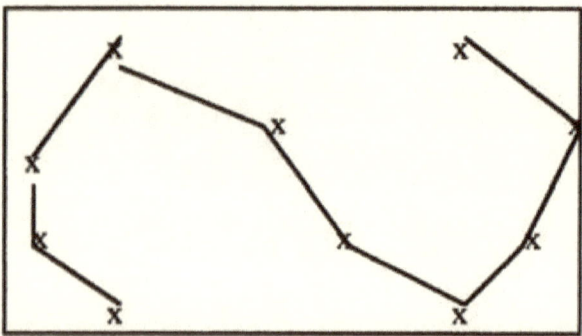

To calculate average depth, total the measurements, and divide by the number of measurements.

A 3.5-4.0 meter long 3/4-inch PVC pipe with a 30-45 cm diameter 1-2 cm thick wooden disc/ metal plate at the bottom makes a good depth measuring device. The pipe's measurements should be graduated in 1 meter increments.

3 ESTIMATING VOLUME (WATER CAPACITY) OF PONDS

For accurate estimation of pond volume in terms of their water holding capacity, following data will be required.

1. Area of the pond (see Chapter-1).
2. Average depth of the pond (see Chapter-2).
3. **Pond Volume = Pond Area x Avg. depth.**
4. Conversion of volume (in cubic meter) into Liter.

*NOTE: Only measure the inside length and inside width of the pond for area or volume estimation (i.e.- the watery margin areas of the pond) but NOT the bundh or dyke length and width.

For Square/ Rectangular Shaped Ponds

If a pond is a perfect square or rectangle, the following formula applies:

$V = l \times w \times h$

Where V = Volume (in cubic meter)

l = length in meter

w = width in meter

h = average depth in meter

Example #3.1. Measurements show a regular rectangular pond to have dimensions of 120 m length and 80 m breadth and 1.5 m depth. What is the volume?

V = l x w x h

= 120 x 80 x 1.5

= 14400cubic meter (m³).

<u>Convert it into litres by the following formula:</u>

Volume (in litres) = Volume (in m³) x 1000

= 14400 x 1000 = 14400000 litres

*NOTE: A 1 m³ unit holds 1000 litres of water.

Example #3.2. An irregular rectangular pond has the following dimensions. What is the volume?

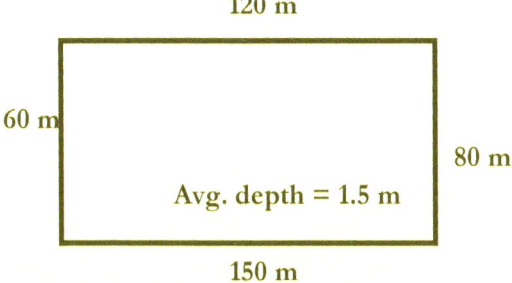

120 m

60 m

80 m

Avg. depth = 1.5 m

150 m

Volume = [(l₁ + l₂)/ 2] x [(w₁ + w₂)/ 2] x h

= [(120 + 150)/ 2] x [(60 + 80)/ 2] x 1.5

= (270/ 2) x (140/ 2) x 1.5

= 135 x 70 x 1.5

= 14175 cubic meter (m³)

<u>Convert it into litres by the following formula:</u>

Volume (in litres) = Volume (in m³) x 1000

= 14175 x 1000 = 14175000 litres

*NOTE: A 1 m³unit holds 1000 litres of water.

For Triangular Shaped Ponds

Calculating pond volume for triangular-shaped ponds is easy. Just use the following formula:

Area = [(a x b)/ 2] x h

Where, a = short side (in meter)

b = long side (in meter)

h = average depth (in meter)

Example #3.3. Calculate the area of a triangular pond with the dimension given below.

a = 80 m

Average depth = 1.5 m

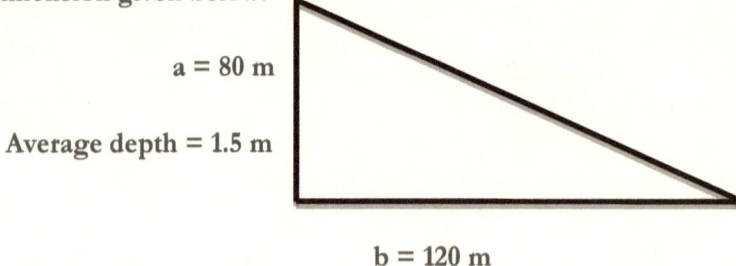

b = 120 m

Area = [(a x b)/ 2] x h
= [(80 x 120)/ 2] x 1.5
= (9600/ 2) x 1.5
= 7200 cubic meter (m^3)
Convert it into litres by the following formula:
Volume (in litres) = Volume (in m^3) x 1000
= 7200 x 1000 = 7200000 litres
*NOTE: A 1 m^3 unit holds 1000 litres of water.

For Trapezoidal Shaped Ponds

To calculate volume for trapezoid-shaped ponds having four sides and a nearly or exactly 90 degree angle, use the following formula:
Volume = [{(a + b) x h}/ 2] x d

Where,

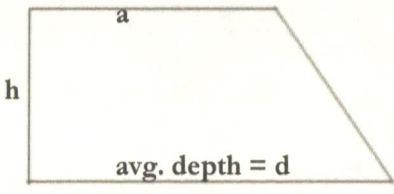

a

h

avg. depth = d

b

Example #3.4. Calculate the area of a trapezoid- shaped pond having the dimensions given in the figure below.

a = 60 m   Avg. depth, d = 1.5 m

h =   40 m

b = 90 m

Area = [{(a + b) x h}/ 2] x d

= [{(60 + 90) x 40}/ 2] x 1.5

= [(150 x 40)/ 2] x 1.5

= (6000/ 2) x 1.5

= 3000 x 1.5

= 4500 cubic meter (m³)

Convert it into litres by the following formula:

Volume (in litres) = Volume (in m³) x 1000

= 4500 x 1000 = 4500000 litres

*NOTE: A 1 m³ unit holds 1000 litres of water.

For Circular Shaped Ponds

If a pond is circular or near circular, the following formula applies:

Volume = [3.14 x (d/ 2)²] x h

Where, d = Diameter of the pond (in meters)

h = Average depth

Example #3.5. Calculate the volume of a circular shaped pond having the diameter of 80 meter and an average depth of 1.5 m.

Avg. depth, h = 1.5 m

Volume = [3.14 x (d/ 2)²] x h

= [3.14 x (80/ 2)²] x 1.5

= 3.14 x 40² x 1.5

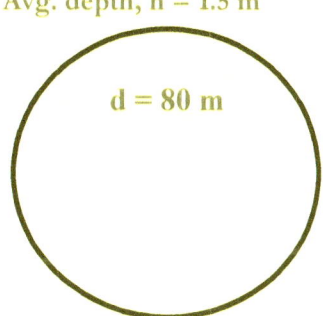

d = 80 m

= 3.14 x 1600 x 1.5

= 7536 cubic meter (m³)

Convert it into litres by the following formula:

Volume (in litres) = Volume (in m³) x 1000

= 7536 x 1000 = 7536000 litres

*NOTE: A 1 m³ unit holds 1000 litres of water.

Irregularly-Shaped Ponds

For calculating volume for irregularly-shaped or mixed- shaped ponds, divide the pond into regular shaped sections and calculate those volumes. Then add the volume of each section to determine the total volume for the pond.

4 CALCULATING TREATMENTS

Basic Treatment Formula

Most chemical treatments can be calculated by using the following formula:

Amount of chemical needed (in grams) = V (in m³) x CF (=1) x ppm desired x [100/ A.I. (in %)]

Where, V= Volume of pond needing treatment (in m³)

CF= Conversion factor (i.e. 1 gm/m³)

ppm = Desired concentration of chemical needed for water volume being treated

A.I. = Active ingredient of the chemical to be used (in %)

*NOTE: Most chemicals are considered 100% AI. The percent AI is the percentage of the active ingredient of the product used and is found on the product's label.

Example #4.1. A pond needs to be treated with an herbicide. The pond is 1.2 hectares or 2.965 acres and has an average depth of 1.5 m. The recommended treatment is 2 ppm. The active ingredient is 100%. How much amount of the herbicide is needed for the treatment?

Given,

Pond area = 1.2 ha or 2.965 acres

Average depth of the pond = 1.5 m

Recommended treatment for herbicide = 2 ppm

A.I. of the herbicide = 100%

I. Convert the area (in hectares/ acres) into area (in square meter) by the following formula:

Area (in square meter) = Area (in hectares) x 10000

= 1.2 x 10000 = 12000 m²

OR, Area (in square meter) = Area (in acres) x 4048

= 2.965 x 4048 = 12002.32 m² ≅ 12000 m²

II. Calculate the volume of the pond by the following formula:

Volume (in cubic meter) = Area (in square meter) x Average depth (in meter)

= 12000 x 1.5 = 18000 m³

III. Apply the treatment formula:

Amount of chemical needed (in grams) = V (in m³) **x CF** (=1) **x ppm desired x [100/ A.I.** (in %)**]**

= 18000 x 1 x 2 x (100/100)

= 18000 x 1 x 2 x 1

= 36000 grams

IV. Convert chemical needed (in grams) into chemical needed (in kilograms) by the following formula:

Chemical needed (in kilograms) = Chemical needed (in grams)/ 1000

= 36000/ 1000 = 36 kg.

CONCLUSION: *Hence, 36 kg of herbicide (with 100% A.I.) will be required to treat the pond having 1.2 ha or 2.965 acres area and 1.5 m average depth, in order to obtain 2 ppm concentration.*

Example #4.2. A pond needs to be treated with Deltamethrin. The pond is 1.2 hectares or 2.965 acres and has an average depth of 1.5 m. The recommended treatment is 1.5 ppm. The active ingredient is 72%. How much amount of the pesticide is needed for the treatment?

Given,

Pond area = 1.2 ha or 2.965 acres

Average depth of the pond = 1.5 m

Recommended treatment for herbicide = 1.5 ppm

A.I. of Deltamethrin (pesticide) = 72%

I. Convert the area (in hectares/ acres) into area (in square meter) by the following formula:

Area (in square meter) = Area (in hectares) x 10000

= 1.2 x 10000 = 12000 m²

OR, Area (in square meter) = Area (in acres) x 4048

= 2.965 x 4048 = 12002.32 m² ≅ 12000 m²

II. Calculate the volume of the pond by the following formula:

Volume (in cubic meter) = Area (in square meter) x Average depth (in meter)

= 12000 x 1.5 = 18000 m³

III. Apply the treatment formula:

Amount of chemical needed (in grams) = V (in m³) **x CF** (=1) **x ppm desired x [100/ A.I.** (in %)**]**

= 18000 x 1 x 1.5 x (100/72)

= 18000 x 1 x 1.5 x 1.39

= 37530 grams

IV. Convert chemical needed (in grams) into chemical needed (in kilograms) by the following formula:

Chemical needed (in kilograms) = Chemical needed (in grams)/ 1000

= 37530/ 1000 = 37.53 kg.

CONCLUSION: *Hence, 37.53 kg of pesticide (with 72% A.I.) will be required to treat the pond having 1.2 ha or 2.965 acres area and 1.5 m average depth, in order to obtain 1.5 ppm concentration.*

*NOTE: Formalin, Salt, Potassium permanganate, Copper sulphate, Methylene blue used for prophylaxis/ therapy in fish farming is CONSIDERED to have 100% A.I.

5 CALCULATING WATER REQUIREMENTS, WATER FILLING TIME AND FLOW RATES

POINT TO REMEMBER: **A 1 m^3 unit holds 1000 litres of water. A 1 hectare-meter pond holds 10000 litres of water. A 1 acre-meter pond holds 4000 litres of water**

For determining minimum water flow rate to fill the pond within a fixed time (t, in days) use the following formula: -

Flow rate (in litres per minute) = Pond volume (in litres)/ [t (in days) x 24 x 60]

Now, convert minutes into days by the following formula:

Filling time (in days) = Filling time (in minutes)/ (60 x 24)

The following examples will illustrate the steps and calculations required to estimate the water requirements for various production units.

Example #5.1. A farmer wants to construct four ponds in the same area and fill all the ponds from one well. The ponds vary in size and the owner wants to be able to fill any pond within7 days. The sizes of the ponds are 6 acres (2.43 ha), 4 acres (1.62 ha), 5.5 (2.23 ha) acres and 2.5acres (1.01 ha). The average water depth in each pond is 5 feet. What flow rate in litres per minute (lpm) is required from the service well to fill any pond in at least 7 days?

    I. First determine the volume of water in the largest pond. If the largest pond can be filled in 7 days, then any smaller ponds will fill in 7 days or less.

Area of largest pond = 6 acres or 2.43 ha.

a) Convert the area (in hectares/ acres) into area (in square meter) by the following formula:

Area (in square meter) = Area (in hectares) x 10000

= 2.43 x 10000 = 24300 m^2

OR, Area (in square meter) = Area (in acres) x 4048

= 6 x 4048 = 24288 m$^2 \cong$ 24300 m^2

*NOTE: 1 hectare (ha) land is equal to 10000 m^2 i.e. - a pond which is 100 m in length and 100 m in breadth. 1 ha area is equivalent to 2.47 acres. 1 acre land is equal to 4048 m^2.

b) Calculate the volume of the pond by the following formula:

Given, average depth required = 5 feet

Converting feet into metre: -

Depth (in metre) = Depth (in feet) x 0.3

= 5 x 0.3 = 1.5 m

*NOTE: 1 feet is equal to 0.30 metre.

Now, Volume (in cubic meter) = Area (in square meter) x Average depth (in meter)

= 24300 x 1.5 = 36450 m^3

II. Convert volume (in cubic metre) into litres:

Volume (in litres) = Volume (in m^3) x 1000

= 36450 x 1000 = 36450000 litres

*NOTE: A 1 m^3 unit holds 1000 litres of water. 1 hectare-meter pond holds 10000 litres of water. 1 acre-meter pond holds 4000 litres of water.

III. Determine the minimum flow rate needed to fill the pond in 7days.

For determining minimum water flow rate to fill the pond(s) within a fixed time (t, in days) use the following formula: -

Flow rate (in litres per minute) = Pond volume (in litres)/ [t (in days) x 24 x 60]

= 36450000/ (7 x 24 x 60)

= 36450000/ 10080

= 3616.07 $\cong$ 3700 litres per minute (lpm)

CONCLUSION: *A flow rate of 3700 lpm should be adequate to fill all the ponds within 7 days.*

Example #5.2. With the 3700 lpm flow rate from Example #5.1,

what would be the filling time in days for the smallest pond of 2.5acres (1.01 ha) and 5 feet average water depth?

I. First determine the volume of water in the pond.

Given, area of the pond = 2.5 acres or 1.01 ha.

a) <u>Convert the area (in hectares/ acres) into area (in square meter) by the following formula</u>:

Area (in square meter) = Area (in hectares) x 10000

= 1.01 x 10000 = 10100 m^2

OR, Area (in square meter) = Area (in acres) x 4048

= 2.5 x 4048 = 10120 m$^2 \cong$ 10100 m^2

*NOTE: 1 hectare (ha) land is equal to 10000 m^2 i.e. - a pond which is 100 m in length and 100 m in breadth. 1 ha area is equivalent to 2.47 acres. 1 acre land is equal to 4048 m^2.

b) <u>Calculate the volume of the pond by the following formula</u>:

Given, average depth required = 5 feet

Converting feet into metre: -

Depth (in metre) = Depth (in feet) x 0.3

= 5 x 0.3 = 1.5 m

*NOTE: 1 foot is equal to 0.30 metre.

Now, Volume (in cubic meter) = Area (in square meter) x Average depth (in meter)

= 10100 x 1.5 = 15150 m^3

II. Convert volume (in cubic metre) into litres:

Volume (in litres) = Volume (in m^3) x 1000

= 15150 x 1000 = 15150000 litres

*NOTE: A 1 m^{3}unit holds 1000 litres of water. 1 hectare-meter pond holds 10000 litres of water. 1 acre-meter pond holds 4000 litres of water

III. Estimating the filling time of the pond at a given flow rate:

Given, flow rate = 3700 lpm (litres per minute)

With a flow rate of Flpm, the filling time (in minutes) can be calculated by the following formula:

Filling time (in minutes) = Pond volume (in litres)/ Flow rate (in lpm)

= 15150000/ 3700

= 4094.594 $\cong$4095 minutes

IV. Convert minutes to days:

Convert minutes into days by the following formula:

Filling time (in days) = Filling time (in minutes)/ (60 x 24)

= 4095/ (60 x 24)

= 4095/ 1440

= 2.84 days ≅3 days

CONCLUSION: *Almost 3 days will be required to fill the 2.5 acres (or 1.01 ha) pond upto a depth of 5 feet at a flow rate of 3700 lpm.*

HOW TO DETERMINE FLOW RATES IN PIPES AT FISH FARM?

Fish culturists often need to adjust the supply of water in a pipe to obtain a desired water exchange or flow rate.

The flow rate through a pipe is easiest to determine by using a container of a known volume. Following steps needs to be followed: -

I. A 1 litre container can be used or a larger container of known volume.

II. Also needed is a stop watch or watch with a second hand.

III. Turn on the water, then place container under pipe to collect water.

IV. With your watch, determine the time that it takes for the container to fill completely.

V. Then make the following calculation:

Flow rate (in lpm) = [Volume of container (in litres) x 60]/ Total seconds to fill container

This procedure should be repeated several times to confirm consistent results and determine an average value.

Example #5.3.A 10 litres container filled up in 1 minute and 20 seconds. What is the flow rate of the supply pipe?

Given, Volume of the container = 10 litres

Time taken to fill container = 1 min 20 secs = 80 seconds

Flow rate (in lpm) = [Volume of container (in litres) x 60]/ Total seconds to fill container

= (10 x 60)/ 80

= 600/ 80

= 7.5 lpm

CONCLUSION: Flow rate of the supply pipe is 7.5 lpm.

6 CALCULATING FISH STOCKS IN A POND

The following situations are related to stocking fish in ponds. Fish stocking rates are usually based on the surface area of water

Example #6.1. A pond is 0.8 hectare in size and 4,500 fish per ha is the desired stocking rate. How many fish are needed?

Given, Pond area = 0.8 ha.

Stocking rate = 4500 fish/ha.

Therefore, applying the following formula:

No. fish seed required = Area of pond (in hectares) x Stocking rate (in fish/ hectare)

0.8hax 4,500 fish/ ha

= 3600 fish

CONCLUSION: *3600 fish seeds will be required to stock the pond at required stocking density.*

Example #6.2. A fish farmer wants to stock 3600 fish in a pond. The fish he wants weigh 120 kg per 1,000 fish. How many kg of fish should be stocked?

Given, No. of fish seed required = 3600

Total sample weight of 1000 fish = 120 kg

No. of fish in the sample, N = 1000

Therefore, applying the following formula:

Fish seed required (in kg) = No. of fish seed required x [Total weight of N no. of fish (in kg)/ N]

= 3600 x (120/ 1000)

$= 3600 \times 0.12$

$= 432$ kg

CONCLUSION: *432 kg fish seed will be required to stock the pond at required stocking density.*

Example #6.3. A fish buyer wants to sample some fish from his pond to check the number of fish that he ordered. He samples 150 fish that weigh 18 kg. The total kg of fish stocked into his pond was 432 kg. How many fishes were stocked?

Given,

Total kg of fish stocked in pond = 432 kg

Total sample weight of 150 fish = 18 kg

No. of fishes in the sample, N = 150

Therefore, applying the following formula:

Number of fish seed stocked = N x [Total weight of fish seed stocked(in kg)/ Total weight of N no. of fish (in kg)]

$= 150 \times (432/ 18)$

$= 150 \times 24$

$= 3600$

CONCLUSION: *3600 fishes were stocked in the pond.*

7 DETERMINING FEED CONVERSION RATIO (FCR)

Feed conversion ratios are calculated to determine the cost and efficiency of feeding. It is affected by the quality of feed, size and condition of fish, number of good feeding days related to temperature and water quality, and feeding practices adopted. Ideally feed conversion ratio should be from 1 to 2.

Use the following procedure to determine the feed conversion ratio for your fish. Keep the following records: -

1. Amount of feed fed daily.

2. Initial weight (in kg) and number of fish stocked.

3. Estimated or final weight (in kg) of standing crop (fish), which is determined by pond sampling.

Now for determining FCR during a particular period, use the following formula: -

FCR = Feed given to fish during the period (in kg)/ [Final fish weight (in kg) – Initial fish weight (in kg)]

POINT TO REMEMBER: Generally under field conditions, 1 million (10 lakh) carp spawn weigh 1.4 kg. 1 lakh carp fry weigh 13 kg. 1000 carp fingerling weigh 15 kg. As a thumb rule, these figures can also be applied for other fishes like catfishes, murrels, perches, etc.

Example #7.1: 8000 fingerlings were stocked in a pond. Initially, 1000 fingerlings weighed 15 kg. Later, the fish were sampled and the

average weight of fish was found to be **45 kg per 1000 fingerling.
During this time 300 kg of feed were fed. What is the feed conversion
ratio?**

I.   Calculating initial weight of fish stocked

Given,

Stocking rate = 8000 fingerlings

Fish seed weight = 15 kg/ 1000 fingerlings

So, total weight of fish seed stocked (initial weight) = 8000 x 15/1000

= 8 x 15 = 120 kg

II.   Calculating final weight of fish stocked

Given,

Stocking rate = 8000 fingerlings

After sampling, average weight of the sample = 45 kg/ 1000 fingerlings

So, total weight of fish seed after sampling (final weight) = 8000 x 45/1000

= 8 x 45 = 360 kg

III.   Calculating FCR

Given,

Amount of feed fed (from stocking upto the day before sampling) = 300 kg

Therefore, applying the following formula:

**FCR = Feed given to fish during the period (in kg)/ [Final fish
weight (in kg) – Initial fish weight (in kg)]**

= 300/ (360-120)

= 300/240 = 1.25

CONCLUSION: FCR is 1.25, i.e. - during the period fish consumed an average of 1.25 kg of feed to gain 1 kg in weight.

*NOTE: Ideally feed conversion ratio should be from 1 to 1.5 and must not exceed 2.

8 CALCULATION OF FEED REQUIREMENT

Recommended feeding rates for different culture practices:
1.   Nursery culture (Spawn to Fry; 15 days culture): -
In first week, give feed @4 times (400%) of the initial body weight/day. In second week, feed @8 times (800%) of the initial body weight/day.
2.   Rearing culture (Fry to Fingerling; 3 months culture): -
In first month, give feed @4% of the initial body weight/day. In second month, feed @6% of the initial body weight/day. In third (last) month, feed @8% of the initial body weight/day.
3.   Grow-out culture (Fingerling to Table sized; 8-11 months): -
Give feed @2-3% of the initial body weight/day.
POINT TO REMEMBER: Generally under field conditions, 1 million (10 lakh) carp spawn weigh 1.4 kg. 1 lakh carp fry weigh 13 kg. 1000 carp fingerling weigh 15 kg. As a thumb rule, these figures can also be applied for other fishes like catfishes, murrels, perches, etc.

Now for calculating feed requirement per day, use the following formula: -
Feed requirement per day (in kg) = [Recommended feeding rate (N) /100] x Total weight of seed stocked in the pond (in kg)
Where, N = Recommended feeding rate (in %)
Now for calculating total feed requirement, use the following formula: -
Total feed requirement (in kg) = Food requirement (in kg) x Number of days

8.1. FEED REQUIREMENT FOR NURSERY CULTURE:

Example #8.1: A farmer stocked 8 million spawn in his nursery pond. His target is to produce fry in 2 weeks. Calculate the daily ration and the amount of feed required during this culture. Also calculate the amount of feed ingredients required, if the feed is to be made of finely powdered rice bran and oil cake @1:1 by weight.

I.   <u>Calculating initial weight of fish seed stocked</u>

Given,

Stocking rate = 8 million

Fish seed weight = 1.4 kg/ 1 million spawn

So, total weight of fish seed stocked (initial weight) = 8 x 1.4/1

= 8 x 1.4 = 11.2 kg

*NOTE - 1 million (10 lakh) carp spawn weigh 1.4 kg.

II.   <u>Calculating feed requirement (FIRST WEEK)</u>

Recommended feeding rate = 4 times or 400% per day

*NOTE: During nursery culture, in first week, give feed @4 times (400%) of the initial body weight per day. In second week, feed @8 times (800%) of the initial body weight per day.

<u>Now for calculating feed requirement per day, use the following formula: -</u>

Feed requirement per day (in kg) = [Recommended feeding rate (N) /100] x Total weight of seed stocked in the pond (in kg)

= [400/100] x 11.2

= 4 x 11.2 = 44.8 kg per day

<u>Now for calculating total feed requirement, use the following formula: -</u>

Total feed requirement (in kg) = Food requirement (in kg) x Number of days

= 44.8 x 7

= 313.6 kg in first week

CONCLUSION: In first week, 44.8 kg feed (22.4 kg rice bran + 22.4 kg oil cake) will be required per day. Total feed requirement is 313.6 kg feed in first week (156.8 kg rice bran + 156.8 kg oil cake).

*NOTE: Feed is to be made of finely powdered rice bran and oil cake @1:1 by weight.

III.   <u>Calculating feed requirement (SECOND WEEK)</u>

Recommended feeding rate = 8 times or 800% per day

*NOTE: During nursery culture, in first week, give feed @4 times

(400%) of the initial body weight per day. In second week, feed @8 times (800%) of the initial body weight per day.

Now for calculating daily feed requirement, use the following formula: -

Feed requirement per day (in kg) = [Recommended feeding rate (N) /100] x Total weight of seed stocked in the pond (in kg)

= [800/100] x 11.2

= 8 x 11.2 = 89.6 kg per day

Now for calculating total feed requirement, use the following formula: -

Daily ration (in kg) = Food requirement (in kg)/ Number of days

= 89.6 x 7 = 627.2 kg

*NOTE: 1 week has 7 days

CONCLUSION: In second week, 89.6 kg feed (44.8 kg rice bran + 44.8 kg oil cake) will be required per day. Total feed requirement is 627.2 kg feed in second week (313.6 kg rice bran + 313.6 kg oil cake).

*NOTE: Feed is to be made of finely powdered rice bran and oil cake @1:1 by weight.

IV. Calculating total feed requirement (during the culture period)

Total feed required = Total feed requirement (first week) + Total feed requirement (second week)

= 313.6 kg + 627.2 kg = 940.8kg

CONCLUSION: The farmer would require a total of 940.8 kg feed (470.4 kg rice bran + 470.4kg oil cake) for growing 8 million spawn in his entire culture period.

*NOTE: Feed is to be made of finely powdered rice bran and oil cake @1:1 by weight.

8.2. FEED REQUIREMENT FOR REARING CULTURE:

Example #8.2: A farmer stocked 2 lakh fry in his rearing pond. His target is to produce fingerling in 3 months. Calculate the daily ration and the amount of feed required during this culture. Also calculate the amount of feed ingredients required, if the feed is to be made of rice bran and oil cake @1:1 by weight.

I. Calculating initial weight of fish seed stocked

Given,

Stocking rate = 2 lakh

Fish seed weight = 13 kg / 1 lakh fry

So, total weight of fish seed stocked (initial weight) = 2 x 13/1

= 2 x 13 = 26 kg

*NOTE - 1 lakh carp fry weigh 13 kg.

II.   Calculating feed requirement (FIRST MONTH)

Recommended feeding rate = 4% per day

*NOTE: In first month, give feed @4% of the initial body weight/day. In second month, feed @6% of the initial body weight/day. In third (last) month, feed @8% of the initial body weight/day.

Now for calculating feed requirement per day, use the following formula: -

Feed requirement per day (in kg) = [Recommended feeding rate (N) /100] x Total weight of seed stocked in the pond (in kg)

= [4/100] x 26

= 0.04 x 26 = 1.04 kg per day

Now for calculating total feed requirement, use the following formula: -

Total feed requirement (in kg) = Food requirement (in kg) x Number of days

= 1.04 x 30

= 31.2 kg

*NOTE: 1 month has 30 days

CONCLUSION: In first month, 1.04 kg feed (0.52 kg rice bran + 0.52 kg oil cake) will be required per day. Total feed requirement is 31.2 kg feed in first month (15.6 kg rice bran + 15.6 kg oil cake).

*NOTE: Feed is to be made of finely powdered rice bran and oil cake @1:1 by weight.

III.   Calculating feed requirement (SECOND MONTH)

Recommended feeding rate = 6% per day

*NOTE: During rearing culture, in first month, give feed @4% of the initial body weight/day. In second month, feed @6% of the initial body weight/day. In third (last) month, feed @8% of the initial body weight/day.

Now for calculating feed requirement per day, use the following formula: -

Feed requirement per day (in kg) = [Recommended feeding rate (N) /100] x Total weight of seed stocked in the pond (in kg)

= [6/100] x 26

= 0.06 x 26 = 1.56 kg per day

Now for calculating total feed requirement, use the following formula: -

Total feed requirement (in kg) = Food requirement (in kg) x Number of days

= 1.56 x 30 = 46.8 kg

*NOTE: 1 month has 30 days

CONCLUSION: In second month, 1.56 kg feed (0.78 kg rice bran + 0.78 kg oil cake) will be required per day. Total feed requirement is 46.8 kg feed (23.4 kg rice bran + 23.4 kg oil cake).

*NOTE: Feed is to be made of finely powdered rice bran and oil cake @1:1 by weight.

IV. Calculating feed requirement (THIRD MONTH)

Recommended feeding rate = 8% per day

*NOTE: During rearing culture, in first month, give feed @4% of the initial body weight/day. In second month, feed @6% of the initial body weight/day. In third (last) month, feed @8% of the initial body weight/day.

Now for calculating feed requirement per day, use the following formula: -

Feed requirement per day (in kg) = [Recommended feeding rate (N) /100] x Total weight of seed stocked in the pond (in kg)

= [8/100] x 26

= 0.08 x 26 = 2.08 kg per day

Now for calculating total feed requirement, use the following formula: -

Total feed requirement (in kg) = Food requirement (in kg) x Number of days

= 2.08 x 30 = 62.4 kg

*NOTE: 1 month has 30 days.

CONCLUSION: In third (last) month, 2.08 kg feed (1.04 kg rice bran + 1.04 kg oil cake) will be required per day. Total feed requirement is 62.4 kg (31.2 kg rice bran + 31.2 kg oil cake).

*NOTE: Feed is to be made of finely powdered rice bran and oil cake @1:1 by weight.

V. Calculating total feed requirement (during the culture period)

Total feed required = Total feed requirement (first month) + Total feed requirement (second month) + Total feed requirement (third month)

= 31.2 kg + 46.8 kg + 62.4 kg = 140.4 kg

CONCLUSION: The farmer would require a total of 140.4 kg feed (70.2 kg rice bran + 70.2 kg oil cake) for growing 2 lakh fry in his entire culture period.

*NOTE: Feed is to be made of finely powdered rice bran and oil cake @1:1 by weight.

8.3. FEED REQUIREMENT FOR GROW-OUT CULTURE

Example #8.3: A farmer stocked 8000 fingerling in his rearing pond. His target is to produce table sized fish in 10 months. Calculate the daily ration and the amount of feed required during this culture. Also calculate the amount of feed ingredients required, if the feed is to be made of rice bran and oil cake @1:1 by weight.

I.   Calculating initial weight of fish seed stocked

Given,

Stocking rate = 8 thousand fingerling

Fish seed weight = 15 kg / 1 thousand fingerling

So, total weight of fish seed stocked (initial weight) = 8 x 15/1

= 8 x 15 = 120 kg

*NOTE - 1000 carp fingerling weigh 15 kg.

II.   Calculating feed requirement (MONTHLY)

Recommended feeding rate = 3% per day

*NOTE: During grow-out culture, give feed @2-3% of the initial body weight/day.

Now for calculating feed requirement per day, use the following formula: -

Feed requirement per day (in kg) = [Recommended feeding rate (N) /100] x Total weight of seed stocked in the pond (in kg)

= [3/100] x 26

= 0.03 x 120 = 3.6 kg per day

Now for calculating total feed requirement, use the following formula: -

Total feed requirement (in kg) = Food requirement (in kg) x Number of days

= 3.6 x 30

= 108 kg per month

*NOTE: 1 month has 30 days

CONCLUSION: In each month, 3.6 kg feed (1.8 kg rice bran + 1.8 kg oil cake) will be required daily. Total monthly feed requirement is 108 kg (54 kg rice bran + 54 kg oil cake).

*NOTE: Feed is to be made of finely powdered rice bran and oil cake @1:1 by weight.

III. Calculating total feed requirement (during the culture period)

Given,

Culture period = 10 months

Total feed required during the culture period = Monthly feed requirement (in kg) x Culture period (in months)

= 108 kg x 10 = 1080 kg

CONCLUSION: The farmer would require a total of 1080 kg feed (540 kg rice bran + 540 kg oil cake) for growing 8 thousand fingerlings in his entire culture period.

*NOTE: Feed is to be made of finely powdered rice bran and oil cake @1:1 by weight.

9 SPECIAL TREATMENTS

In this chapter we will only learn about the application of some special chemicals used in fish culture operation *viz.*- Copper sulphate ($CuSO_4$), Formaldehyde or Formalin (HCHO) and Potassium permanganate ($KMnO_4$). Special care is required during their application or treatment calculation.

9.1. COPPER SULPHATE TREATMENT

Find out the total alkalinity value (in ppm) of pond water. Divide the total alkalinity (in ppm) of pond water by 100 to get the required concentration of copper sulphate (in ppm).

Now for calculating $CuSO_4$ requirement (in ppm), use the following formula: -

Copper sulphate required (in ppm) = Total Alkalinity (in ppm)/ 100

POINT TO REMEMBER: However in any case the treatment of Copper sulphate should not exceed 2 ppm in ponds. Copper sulphate has 100% A.I.

Example #9.1.A pond needs to be treated with Copper sulphate. The pond is 1.2hectaresor 2.965 acres and has an average depth of 1.5 m. The total alkalinity of pond water is 200 ppm. How much amount of copper sulphate is needed for the treatment?

Given,

Total alkalinity of pond water = 200 ppm

Now for calculating $CuSO_4$ requirement (in ppm), use the following formula: -

Copper sulphate required (in ppm) = Total Alkalinity (in ppm)/ 100

= 200/ 100

= 2 ppm

So, a treatment of 2 ppm Copper sulphate will be required in the pond.

Follow **Example #4.1** of **Chapter – 4** for determining the amount of Copper sulphate needed (in grams/ kg) to treat this particular pond in order to obtain 2 ppm concentration.

*NOTE: Copper sulphate has 100% A.I.

9.2. FORMALDEHYDE (FORMALIN) TREATMENT

Treatment of formalin is temperature dependent. Low concentration of formalin is required at higher temperatures. High concentration of formalin is required at lower temperatures. Aeration should always be given in pond or tanks when formalin is applied.

POINT TO REMEMBER: Formalin has 100% A.I. As a thumb rule, 1 ml formalin is equivalent to 1 mg of formalin.

1. For pond treatment, use formalin @15-25 ppm depending on temperature.
2. For bath treatment (less than 60 minutes), use formalin @150-250 ppm depending upon temperature.

Follow **Example #4.1** of **Chapter – 4** for determining the amount of Formalin needed (in gm/kg) to treat a pond in order to obtain a desired concentration.

*NOTE: Generally 1 ml formalin is equivalent to 1 mg in weight. Formalin has 100% A.I.

9.3. POTASSIUM PERMANGANATE TREATMENT

Treatment of Potassium permanganate is dependent on organic load present in water. As a thumb rule, add 2-4 ppm (mg per litre) Potassium permanganate in a pond.

To determine the concentration (in ppm) of $KMnO_4$ needed in a particular pond, a "Demand test" is required. It is given as follows: -

1. Take 10 litres of pond water in a container.
2. Add 2 ppm (i.e. – 2 mg per litre) Potassium permanganate into it and wait.
3. If the purple colour persists for 4 hours, then the pond needs 2 ppm $KMnO_4$. Stop testing.
4. However if the purple colour changes to brownish in less than 4 hours, then further add 2 ppm (i.e. – 2 mg per litre) Potassium permanganate into it and again wait.

5. If the purple colour persists for 4 hours, then the pond needs 4 ppm (2 ppm + 2 ppm) KMnO$_4$. Stop testing.

6. Again, if the purple colour changes to brownish in less than 4 hours, further add 2 ppm (i.e. – 2 mg per litre) Potassium permanganate into it and again wait.

7. If the purple colour persists for 4 hours, then the pond needs 6 ppm (2 ppm + 2 ppm + 2 ppm) KMnO$_4$. Stop testing.

8. Again, if the purple colour changes to brownish in less than 4 hours. Stop testing. The pond is unsuitable for Potassium permanganate application. Reduce the organic load of pond in such case by dewatering, desilting and liming the pond.

POINT TO REMEMBER: Do not add Potassium permanganate in a pond greater than 6 ppm. KMnO$_4$ has 100% A.I.

*NOTE: As a thumb rule, add 2-4 ppm (mg per litre) Potassium permanganate in a pond.

Follow **Example #4.1** of **Chapter – 4** for determining the amount of Potassium permanganate needed (in gm/kg) to treat a pond in order to obtain a desired concentration.

*NOTE: KMnO$_4$ has 100% A.I.

10 CONVERSION CHARTS

Table 1 – Conversion in Volume

From	To						
	cubic inches (in³)	cubic feet (ft³)	fluid ounce (fl oz)	gallon (gal)	cubic centimeter (cm³)	liter (l)	cubic meter (m³)
cubic inches (in³)	1	0.000579	0.5541	0.00433	16.39	0.0164	0.00001
cubic feet (ft³)	1728	1	9575	7.481	0.000283	28.32	0.0283
fluid ounce (fl oz)	1.805	0.00104	1	0.0078	29.57	0.0296	0.00002
gallon (gal)	231	0.1337	128	1	3785	3.785	0.0038
cubic centimeter (cm³)	0.061	0.0000353	0.0338	0.000264	1	0.001	0.000001
liter (l)	60.98	0.0353	33.81	0.2642	1000	1	0.001
cubic meter (m³)	610000	5.31	33800	264.2	1000000	1000	1

*NOTE: cubic centimeter (cm³) = milliliter (ml).

Table 2 – Conversion in Weight

From	To			
	ounce (oz)	pound (lb)	gram (g)	kilogram (kg)
ounce (oz)	1	0.0625	28.35	0.0284
pound (lb)	16	1	453.6	0.4536
gram (g)	0.0353	0.0022	1	0.001
kilogram (kg)	35.27	2.205	1000	1

Table 3 – Conversion in Length

From	To				
	inches (in)	feet (ft)	yard (yd)	centimeter (cm)	meter (m)
inches (in)	1	0.0833	0.0278	2.540	0.0254
feet (ft)	12	1	0.3333	30.48	0.3048
yard (yd)	36	3	1	91.44	0.9144
centimeter (cm)	0.3937	0.0328	0.0109	1	100
meter (m)	39.37	3.281	1.0936	100	1

Table 4 – Conversion of various volumes to attain one part per million (ppm)

Amount Active Ingredient	Unit of Volume	Parts per million
2.71 pounds	acre-feet	1 ppm
1.235 grams	acre-feet	1 ppm
1.24 kilograms	acre-feet	1 ppm
0.0283 grams	cubic feet	1 ppm
1 milligram	liter	1 ppm
8.34 pounds	million gallons	1 ppm
1 gram	cubic meter	1 ppm
0.0038 grams	gallon	1 ppm
3.8 grams	thousand gallons	1 ppm

Box 1 – Relationship between percent (gm/100 ml), parts per thousand (gm/l) and parts per million (mg/l) concentration solution

- ➤ 1 percent solution = 10 ppt solution
- ➤ 1 percent solution = 10000 ppm solution
- ➤ 1 ppt solution = 1000 ppm solution

Box 2 – Miscellaneous conversion factors

- ➤ 1 acre-foot= 43,560 cubic feet
- ➤ 1 acre-foot= 325,850 gallons
- ➤ 1 acre-foot of water= 2,718,144 pounds
- ➤ 1 cubic foot of water= 62.4pounds
- ➤ 1 gallon of water= 8.34 pounds
- ➤ 1 gallon of water= 3,785 grams
- ➤ 1 liter of water= 1,000 grams
- ➤ 1 fluid ounce= 29.57 grams
- ➤ 1 fluid ounce= 1.043 ounces
- ➤ Centigrade to Fahrenheit = $(C° \times 9/5) + 32°$
- ➤ Fahrenheit to Centigrade = $(F° - 32°) \times 5/9$

SUGGESTED READINGS

Dorman, L. 1977. Aquaculture producer's quick reference handbook. Cooperative Extension Program, University of Arkansas. Pub No.MP435-1M-1-07RV.

Jensen, G. 1998. Handbook for common calculations in fin fish aquaculture. Louisiana Agricultural Experiment Station, Louisiana Cooperative Extension Service. Pub No. 8903 (500). http://www.agctr.lsu.edu/wwwac

Piper, R. G., McElwain, I. B., Orme, L. E., McCraren, J. P., Fowler, L. G. and Leonard, J. R. 1982. Fish Hatchery Management. U.S. Department of Interior, Fish and Wildlife Service, Washington, D.C.

United States Fish and Wildlife Service. 1981. Diseases of Hatchery Fish. Superintendent of Documents, U.S. Government Printing Office.

USDA/Soil Conservation Service. 1982. Aquaculture pumping plants for water management. Engineering Technical Note.

Wellborn, T. L. 1987. Catfish Farmer's Handbook. Extension Wildlife and Fisheries Department, Mississippi Cooperative Extension Service, Mississippi State University.

ABOUT THE AUTHOR

Koushik Roy hails from Haripal, Hooghly, West Bengal (India). He is now a Senior Research Fellow (NICRA) in ICAR - Central Inland Fisheries Research Institute, Barrackpore, Government of India. He was also an ex-guest faculty in department of Industrial Aquaculture and Fisheries (B.Voc.), Asutosh College, University of Calcutta. He Completed M.F.Sc. in Aquaculture (1st class 1st) from Indira Gandhi Krishi Vishwavidyalaya in 2014 and B.Sc. in Industrial Fish and Fisheries (1st class 1st) from University of Calcutta on 2011. He has qualified ICAR-JRF (2011), SRF (2015), NET and ARS (Mains) in 2014. Koushik Roy did his M.F.Sc. Thesis work under the able guidance of Dr. Chari and Dr. Gaur.

http://in.linkedin.com/in/koushikroy89

The co-authors, Dr. M.S. Chari and Dr. S.R. Gaur (Head) are Professors in Dept. of Fisheries, Indira Gandhi Krishi Vishwavidyalaya, Raipur, Chhattishgarh, India. They individually have 20+ years of combined teaching, research and administrative experience in the field of fisheries and aquaculture. They have authored numerous national and international publications. They also acted as reviewers in reputed journals, ex-member in various task forces of the government for fisheries. They had visited as delegates to various national and international conferences/ seminars and appeared in state level TV shows, radio talks for the dissemination and extension of knowledge, which was their primary aim throughout the long and glamorous career.

www.ingramcontent.com/pod-product-compliance
Lightning Source LLC
Chambersburg PA
CBHW021444170526
45164CB00001B/382